Ann Weil

www.steck-vaughn.com

Photography: Cover courtesy of NASA; pp.iii (background), p.iv ©NASA; p.2 ©Bettmann/CORBIS; pp.6–7, 11, 15, 18, 20–21, 23, 25, 27, 30, 33, 35, 36, 38, ©NASA; p.40 ©Roger Ressmeyer/CORBIS; pp.41, 45–47 ©NASA; p.50 ©Reuters NewMedia Inc./CORBIS; pp.56–60 ©NASA.

Additional photography by CORBIS and Corel Corporation.

ISBN 0-7398-5173-X

Printed in the United States of America.

1 2 3 4 5 6 7 8 9 LB 06 05 04 03 02

Contents

CHAPTER 1
The Race to Space

A rocket taking off for outer space is a common event these days. However, space travel was just **science fiction** only fifty or sixty years ago. It took competition between two countries to turn science fiction into fact.

The Cold War

In the 1950s, the United States and the Soviet Union were in a power struggle. (Present-day Russia was part of the Soviet Union.) This time was known as the Cold War. The two countries were not really at war, but they were very suspicious of one another. Each country wanted to be the first to explore space.

◀ **In 1969, *Apollo 11* took off for the moon.**

The Soviets Were First

On October 4, 1957, the Soviets launched *Sputnik 1. Sputnik* means "satellite" in Russian. *Sputnik 1* was the first satellite successfully launched into orbit around Earth.

One month later, the Soviets sent a dog into space on board *Sputnik 2. Laika*, which means "barker" in Russian, was the first living creature sent into space.

The Soviets then had another **significant** first. On April 12, 1961, Yuri Gagarin made a full orbit around Earth on board a Soviet rocket.

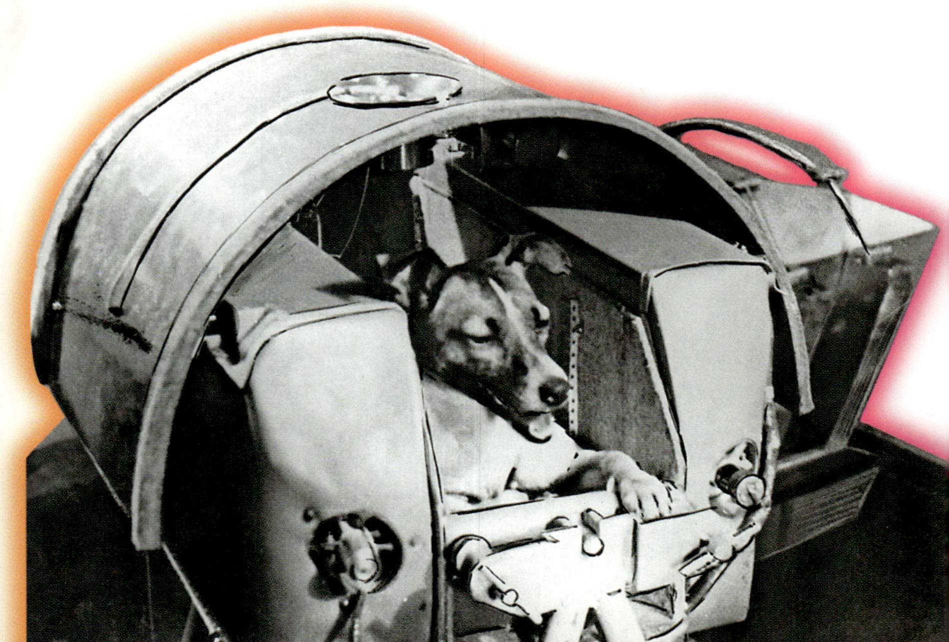

Man on the Moon

The United States wanted to catch up to and **overtake** the Soviet Union in the space race. This competition was driven by more than national pride. The U. S. government worried that the Soviets might use their space **technology** as a weapon against the American people.

The National Aeronautics and Space Administration (NASA) was created in October 1958. Its official mission was to plan and carry out space activities. Its real mission was to beat the Soviets to the moon.

In 1961, President John F. Kennedy told the American people he wanted to send astronauts to the moon and then return them safely home. It took almost ten years to do it. On July 20, 1969, American astronaut Neil Armstrong became the first person to set foot on the moon. When he climbed down the ladder onto the moon, Armstrong said, "That's one small step for a man, one giant leap for **mankind**."

Laika was the first Earth creature to take a trip to outer space.

Armstrong and another *Apollo 11* astronaut, Edwin "Buzz" Aldrin, Jr., walked on the moon. Millions of people around the world watched them on television. The two men planted an American flag on the moon. That showed the world that the United States was the new leader in space.

This **achievement** captured people's imaginations. Suddenly, space exploration held all kinds of wonderful possibilities.

This famous moment was beamed into living rooms around the world.

First Space Stations

The first space stations were small laboratories that orbited Earth.

Salyut

Salyut 1 was the world's first space station. The Soviet Union launched it on April 19, 1971. The first crew went by rocket to the space station three days later, but they could not open *Salyut's* hatch. They returned to Earth without having entered the space station.

The second *Salyut* crew got into the space station and stayed for 24 days. As the three **cosmonauts** returned to Earth, a valve opened and the air inside the spacecraft escaped. The men were not wearing space suits to help them breathe. They died from a lack of oxygen. After that accident, all cosmonauts wore space suits when traveling to and from the space station.

Skylab

The U. S. space station, *Skylab*, was launched in May 1973. It was more than three times larger than *Salyut*. It weighed about one hundred tons. *Skylab* was damaged as it went into orbit. NASA sent a team of three astronauts to fix the space station.

The astronauts made the repairs while *Skylab* orbited Earth. This mission showed that **complex** construction tasks could be done in space. *Skylab* missions later proved that humans could live and work in space for months at a time.

Mir

The Soviets launched *Mir* in February 1986. *Mir* means "peace" in Russian. *Mir* cosmonauts became the first humans to spend more than a year in space.

Mir was the first space station designed to be **expandable.** It was made up of different pieces called **modules.** Over time, new modules were added to the space station. This gave the cosmonauts more room to live and work.

The early space stations *Salyut*, *Skylab*, and *Mir* are no longer in orbit.

Skylab **blasted into space with no astronauts aboard.**

From Competition to Cooperation

"Men who have worked together to reach the stars are not likely to descend together into the depths of war . . . ." These words were spoken by Lyndon B. Johnson in a speech to the United Nations General Assembly. At that time he was a United States Senator. Later, as John F. Kennedy's Vice President, Johnson was in charge of America's first space program.

In the early 1990s, the Soviet Union broke up into many different countries. One of them was Russia. With the end of the Cold War, the United States and Russia began working together on Earth and in outer space.

Instead of keeping up their own separate space stations, the two countries worked together to build and launch the *International Space Station* (*ISS*). With the *ISS*, Johnson's dream of international **cooperation** finally became a reality.

The Shuttle-Mir Program

The *ISS* needed years of planning. During that time, astronauts from America and other countries lived and worked on board *Mir*. The Shuttle-Mir program began in 1995. It included nine docking missions between the United States **space shuttle** and the Russian space station.

This program was a great success. It taught the world's scientists a lot about international space operations and space technology.

The *International Space Station*

The *ISS* is the largest scientific project ever. It shows international cooperation at its best. The United States, Russia, Japan, Canada, Brazil, and 11 European countries share responsibility for the *ISS*.

Research on the *ISS*

Every *ISS* mission includes many different scientific experiments. The experiments done in this orbiting laboratory can help us learn more about life in space and life on Earth.

Many experiments look at how being weightless affects humans, plants, and animals. The health of the astronauts is carefully monitored. This information helps us understand the long-term effects that space travel might have on humans.

Solar research carried out on the *ISS* helps us understand the effects of the sun on Earth and in space. The space station uses solar energy to power its equipment. However, the sun's **radiation** can harm the astronauts if they are not well protected.

The *ISS* is also a factory in outer space. Weightless conditions in outer space might help us to **manufacture** new materials for use on Earth. For example, crystals grown in a weightless environment are more perfect than those formed on Earth.

There are many different types of crystals. Some crystals are made of **proteins**. Those crystals grown in the weightless environment of the *ISS* may be used to make better medicines.

Watching the Earth from Outer Space

The *ISS* has a good view of planet Earth. Astronauts on board the space station can observe 85 percent of Earth's surface.

Images and information from the *ISS* help us **evaluate** the effects of air and water pollution. These observations help scientists understand changes in our environment.

Getting to the *ISS*

Astronauts from different countries work together on board the space station. They get to the *ISS* on either a Russian rocket or a United States space shuttle.

This is what Earth's Adriatic Sea looks like from the *ISS*.

Soyuz

The first *ISS* crew traveled to the space station on board a Russian Soyuz rocket. The first *ISS* crew was made up of one American and two Russians. They flew to the *ISS* in October 2000.

The Russians have been using Soyuz rockets for almost thirty years. *Soyuz* means "union" in Russian. This type of rocket is very small. Each one can be used for only one trip to the *ISS* and back. Every time the Russians send a crew up to the *ISS*, they must use a new rocket.

The Space Shuttle

The space shuttle is the world's first **reusable** spacecraft. It's also the first spacecraft that can carry very large, heavy items. It carries satellites and parts of the *International Space Station.*

A space shuttle launches like a rocket, orbits Earth like a spacecraft, and lands like an airplane. It takes the space shuttle about ten minutes to get into its orbit between 115 and 250 miles above Earth.

Once in orbit, a space shuttle can travel at a speed of 17,322 miles per hour. A person traveling that fast could get from New York City to Los Angeles in nine minutes!

The first space shuttle, *Columbia*, lifted off on April 12, 1981. It carried a crew of two astronauts. Today, NASA's fleet of space shuttles also includes *Discovery, Atlantis,* and *Endeavour.*

There was another space shuttle, named *Challenger*. In January 1986, *Challenger* exploded about a minute after taking off. It was destroyed, and everyone on board died.

The *Hubble Space Telescope*

Space shuttle *Discovery* put the *Hubble Space Telescope* into orbit on April 25, 1990. *Hubble* is one of the largest satellites ever built. It is as big as a school bus and weighs 12 tons.

A Few Comparisons

Object	Length	Weight
ISS	88 meters	453, 592 kilograms
Shuttle	56 meters	2,041,000 kilograms
Skylab	36 meters	76,295 kilograms
Mir	13 meters	20,900 kilograms
Salyut	13 meters	18,210 kilograms
Hubble	13 meters	11,000 kilograms
School bus	10.7 meters	10,896 kilograms

For hundreds of years, people have used telescopes to look at the night sky. A space telescope can see farther and more clearly than the most powerful telescopes used on Earth. That's because rain, snow, clouds, air pollution, and city lights do not block the view from the *Hubble Space Telescope.*

The *Hubble Space Telescope* captured this image of a circle of star clusters.

Hubble has given us detailed pictures of Earth, stars, and other planets. The images from the *Hubble Space Telescope* can teach scientists about our universe. For example, *Hubble* photographed the birth of more than one hundred star clusters. We now know that some of these star clusters formed when galaxies smashed into each other, tossing stars and gas into space. Each cluster can contain as many as a million stars.

Astronomers cannot just take a look through *Hubble's* lens to see the universe. Instead, they look at images sent to Earth from *Hubble's* video cameras.

A space shuttle visits the *Hubble Space Telescope* every few years. Astronauts add new equipment and make any needed repairs. Maintaining the *Hubble Space Telescope* and other satellites is an important job for the space shuttle and its crew. Their most important job, though, is building the *International Space Station.*

Building the *International Space Station*

The completed space station will weigh about one million pounds, or as much as five *Skylabs*! It will be about as long as a football field.

It would be impossible to launch something so big and heavy from Earth. That is why the *ISS* is being built in outer space, piece by piece. Space station modules, building materials, and equipment are launched into space. In space, astronauts put together pieces of the space station.

The Russians launched the first *ISS* module, *Zarya*, on November 20, 1998. *Zarya* means "sunrise" in Russian. An American module, named *Unity*, was launched two weeks later. The crew of the space shuttle connected the two modules in orbit. More modules have been added since then.

Early *ISS* crews were limited to three people because of the small quarters. When the *ISS* is completed, it will have as much living space as there is inside three average American homes. An international crew of seven men and women will stay on board the *ISS* for three to six months at a time.

Zarya (left) and _Unity_ (right) were the first two parts of the _International Space Station_.

Astronauts build some of the space station using mechanical robot arms. Crew members operate the robot arms from inside the spacecraft. This is safer and more comfortable for the astronauts than working outside the *ISS*.

Astronauts must wear space suits when they work outside the space station. Outside the *ISS*, there is no air to breathe. Astronauts can breathe normally inside the *ISS*.

Space Walks

Sometimes astronauts must work outside the spacecraft. They may be building some part of the space station or making a repair that cannot be done using the robot arms. The official term for this job is Extra-Vehicular Activity (EVA). This means the astronauts are doing something outside the *ISS* or another spacecraft. Sometimes this is called a space walk.

During a space walk, astronauts stay connected to the spacecraft. If they accidentally break loose, they can use their jet pack "life jacket" to push them back to the spacecraft.

It is very hard to work in a space suit. The gloves astronauts must wear are very bulky. This makes it hard to hold on to things. If an astronaut accidentally drops something, it could just float away and get lost. Special tools were made so astronauts can do their work safely and more easily.

Observing the *ISS* in Your Sky

The *ISS* completes an orbit around Earth about every ninety minutes. People all over the world can see the *ISS* at different times.

The *ISS* looks like a moving star when seen from Earth. It moves quite quickly and crosses the sky in just a few minutes. You can look for the *ISS* on a clear night, away from bright city lights.

An astronaut performs a space walk from the space shuttle *Discovery*.

NASA's Web site can help you find the *International Space Station* in the night sky. You can search for NASA and other helpful Web sites on the Internet. The Web sites can tell you when and where in the sky to look.

Astronaut Selection and Training

NASA chose the first seven astronauts in 1959. All seven men were **test pilots** and they all worked in the United States Armed Forces.

Things have changed a lot since then. NASA still trains pilots, but astronauts come from many different backgrounds now. Some astronauts have a **military** background. Others are scientists who do experiments in outer space.

Carlos Noriega was in elementary school when Neil Armstrong stepped onto the moon. He remembers that time very well.

Astronaut Carlos Noriega thinks his job is a dream come true.

"I was a young kid, just arrived in this country, barely spoke English, and thought, 'well, that's a dream . . . not something I could do because those astronauts must be raised in a palace somewhere, or some laboratory,'" he said.

Noriega thought being an astronaut wasn't for an average kid whose dad worked two or three jobs just to make ends meet. He just put the idea of becoming an astronaut out of his mind.

Years later, Noriega was a United States Marine. He saw someone in his office filling out an application to be an astronaut. Noriega looked at the application. He realized that he himself qualified, so he sent in his own application.

Carlos Noriega was accepted. He loves being an astronaut and has what he calls "the best job in the world."

Training

Most astronauts train for years before they go up into space. There's a lot to learn. NASA uses many training methods to prepare astronauts to do their job in space.

Astronauts train at Johnson Space Center in Houston, Texas. They must complete one year of basic training before they earn the title *astronaut.* They go to classes in science, math, and other subjects.

Some American astronauts must also learn the Russian language. They must be able to speak Russian if they will fly on a Russian spacecraft or work with Russian cosmonauts.

All astronauts learn water and wilderness survival skills. These skills could save their lives if they need to make an emergency landing on Earth.

Just Like the Real Thing

Astronauts practice their missions in a **simulator**. Simulators have the same hand controls as the actual spacecraft. What the astronauts "see" out the window looks exactly like what they will see when they get to outer space.

NASA wants to be sure that astronauts are ready for every kind of emergency. In training, everything that might go wrong does go wrong. Astronauts practice fixing broken equipment and solving problems.

Ellen Ochoa's first mission was on board space shuttle *Discovery* in 1993. She found the training harder than the real missions.

Like all astronauts, Ellen Ochoa trained hard before going into space.

Training Without Gravity

It would be nice if NASA had a room where they could just flip a switch and "turn off" gravity. There isn't any room like that, though. Astronauts in training face the same laws of gravity as everyone else on Earth.

Before going into space, astronauts need to know how to live and work in a weightless environment. NASA uses some creative ways to get astronauts ready for weightless conditions. These solutions don't feel exactly like being in space. Still, the training gives astronauts a good idea of what to expect.

NASA trains astronauts underwater. They sink a **full-sized** model of an *International Space Station* module inside a giant pool. Like a sunken ship, the module is full of water. The astronauts wear a space suit in the water. The space suit weighs about 240 pounds. The astronauts also wear weights and floats so they neither sink nor rise. The astronauts feel weightless wearing their

This underwater environment helps astronauts and cosmonauts prepare to work in outer space.

gear underwater. These conditions help astronauts practice the work they will do while they are on the space shuttle and the *ISS*.

Astronauts use their hands and arms to pull themselves from place to place. There are rails throughout the space shuttle and the *ISS*. Astronaut John "Danny" Olivas found the training experience most interesting. He was surprised at how little astronauts use their legs in the underwater environment.

Underwater training is useful, but it is not exactly like the real thing. Ellen Ochoa found that being weightless feels more natural than being underwater. She thinks being weightless is the most fun part of her job.

The "Vomit Comet"

NASA also trains astronauts in the air. They adapted a jet cargo plane to train astronauts in **weightlessness**. They took out all the seats and padded the whole inside of the plane.

Flying in this plane is like riding a huge **roller coaster**. As the plane plunges toward Earth, the people inside feel weightless. It seems as though they are rising off the floor of the plane. Actually, the plane is falling away beneath them. This condition lasts for about twenty seconds.

Sometimes astronauts practice putting on a space suit while weightless. Twenty seconds is not long enough for astronauts to finish most training activities, so they may take as many as forty flights in just one day!

All astronauts train on this plane. It is nicknamed the "vomit comet," because many astronauts feel dizzy and sick to their stomachs during this wild ride.

Astronauts Training Astronauts

Astronauts in training can learn a lot from other astronauts who have already been in space. Astronaut Robert Curbeam found this very helpful when he was training for his first space walk. "I've talked to probably fifty or sixty percent of the people who've done space walks in the past about what I'm going to do and what I should look out for. And those people have been absolutely great," he said.

"Every day is different," adds Curbeam. "The great thing about it is you know every day is just kind of narrowing your focus into that one day when you're going to get to ride the rocket."

OVHD
EXIT
AV-1

CHAPTER 4

On the Job in Outer Space

3…2…1…Lift off! The months of waiting and years of training all lead up to this moment. Astronauts have a big job to do, and working in space isn't your average day at the office.

No Up or Down

The *ISS* has equipment everywhere—on the walls, on the floors, and on the ceilings. In fact, there really is no floor or ceiling in space. There is no real up or down. Because the astronauts are weightless, they feel the same no matter which way they are placed. Because of that, it can be easier to work in space than on Earth. Working upside down is just as easy as working right side up.

Astronaut Robert Curbeam installs new equipment on the *ISS*.

Space Sickness

Most astronauts experience some kind of space sickness. Without gravity, fluids inside our bodies move differently than on Earth. This feels strange and uncomfortable.

Astronaut Mae Jemison was part of a motion sickness study while orbiting Earth in the space shuttle. Her first trip into space was in 1992 on board space shuttle *Endeavour*.

Many astronauts take medicine when they feel space sick. Instead, Jemison used **biofeedback** to combat her motion sickness. She relaxed her body to fight the **nausea** without medicine. This was part of an experiment on space sickness.

Space sickness goes away when the body gets used to life in space. Most astronauts feel better during their second and third flights.

Instruments check Mae Jemison's physical responses to space travel.

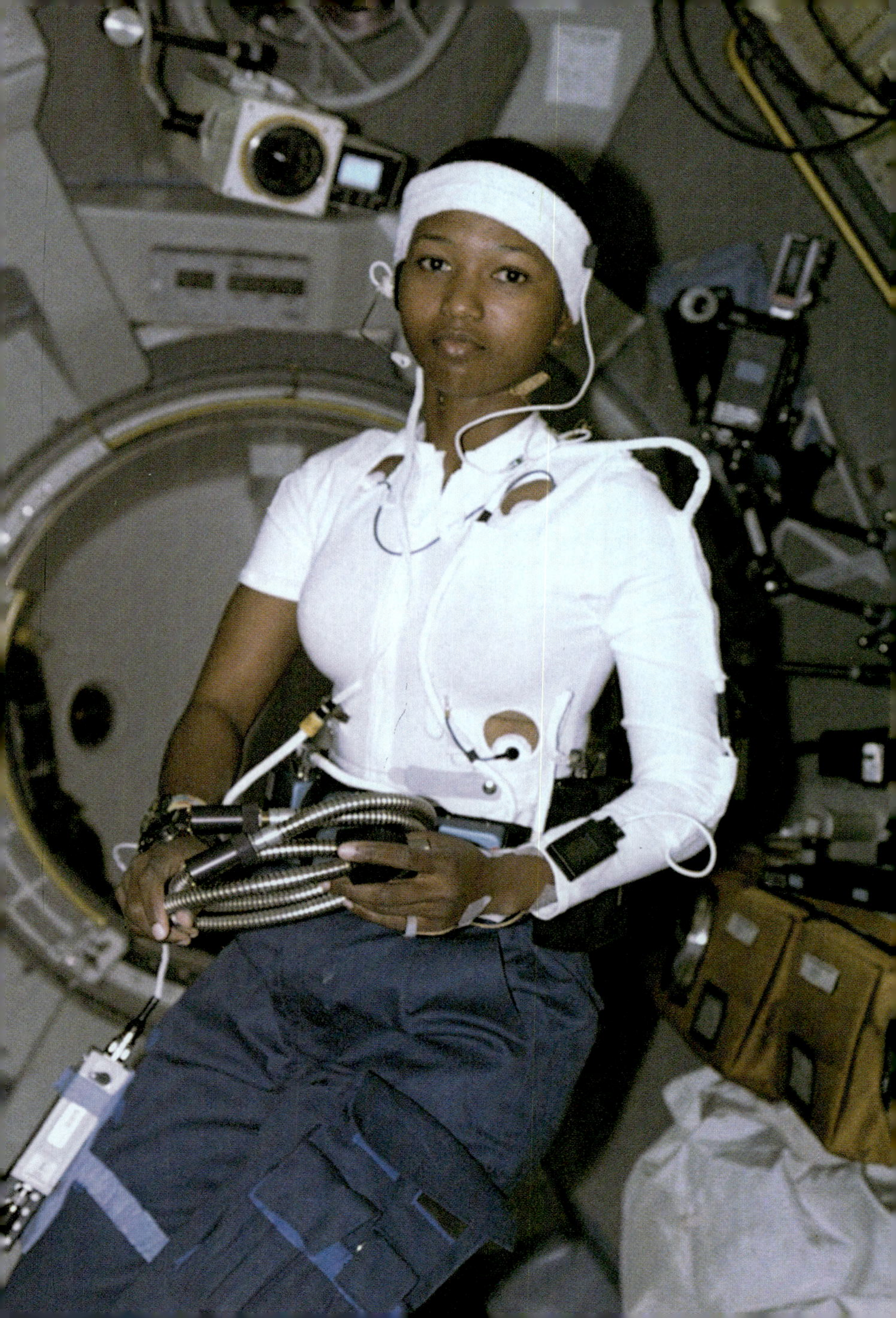

Waking Up in Outer Space

Astronauts in space wake up, have breakfast, and "go to work" just like millions of people on Earth. As they orbit the Earth, the sun seems to rise and set many times during a 24-hour period. Astronauts in space can't count on the sun to wake them up.

This doesn't mean that astronauts get to sleep late. NASA mission control plays loud music inside the *ISS* when it's time for the astronauts to start their day.

Working in Outer Space

Shuttle and *ISS* astronauts work on experiments in many different fields of science. Some experiments test how animals and humans adapt to being weightless.

One of space shuttle *Challenger's* 1985 missions studied how animals behaved in space. Two monkeys and 24 rats were on board.

"We didn't take them out of their cages," shuttle pilot Frederick Gregory said. "Part of the experiment was to try out the cages that were being used to make sure they provided a good environment for the animals."

Gregory was able to see how the two monkeys reacted to being weightless. Like people, the monkeys adapted at their own pace. One got used to it right away. The other took a day and a half.

Astronaut Frederick Gregory observed animals in a weightless environment.

Eating in Space

Eating may seem like a simple thing to do. After all, we all do it several times a day. Learning to eat in space can be quite challenging.

Here on Earth you can just sit at a table with a plate of food in front of you. In space that plate would float around, and the food would get all

over the place. It would be pretty messy. It could also be dangerous. Food crumbs could float into machinery. Liquid from a spilled drink could ruin delicate equipment.

Astronauts need to eat very carefully. They use trays with straps that tie around their laps. The trays have **magnets** so the forks and spoons don't float away. Even without gravity, foods like pudding stick to a spoon!

When astronauts are thirsty, they can't just pop open a can or bottle and take a drink. The liquid would float out into the air. Instead, astronauts add water to a packet filled with a powdered drink. Then they drink it with a straw. After they take a sip, they have to pinch the straw, or else the liquid will escape.

A lot of space food is **dehydrated**. This means some of the water has been taken out of the food. Dehydrated food weighs less and takes up less space. Astronauts add water to the dehydrated food before eating it. It still doesn't taste the same as regular food.

Scott Horowitz is making a peanut butter sandwich that is out of this world!

Some foods are the same you would find at your local supermarket. Nuts, candy, pudding, dried fruits, and peanut butter are all used in outer space. The *ISS* has a refrigerator and a **microwave oven**.

Working Out in Space

Without the pull of gravity, getting around the space station is pretty easy. You can launch yourself across a room with just the flick of your finger against a wall. You can stop yourself by grabbing onto something. It's easy to move something heavy when there is no force of gravity holding it down.

Everyone seems very strong in space, but they are not using much muscle power. If they don't exercise, astronauts can become weak. Daily workouts keep an astronaut's body strong.

Washing in Space

Do you like to take a long, hot shower when you wake up in the morning? Maybe you like bubble baths. Astronauts do not have such baths or showers. Without gravity, water would not run down your body when you shower. Bath water would float around you without actually touching your skin. The water might even go up your nose!

Astronauts take sponge baths or use wet wipes to keep themselves clean. They use a special shampoo that they comb through their hair. Then the astronauts wipe the shampoo and dirt from their hair using a towel.

Astronauts brush their teeth pretty much the same way they do on Earth. They spit into a tissue instead of using a sink, though.

How *Do* You Go to the Bathroom in Space?

Every astronaut hears this question at least once! NASA made a special toilet for use in outer space. It looks a little bit like a toilet on an airplane, but it doesn't flush. Solid waste is wrapped in plastic. Liquid waste is vacuumed away from the body using a **funnel**.

Astronauts on the *ISS* use a bathroom like this one.

A large strap holds Mission Specialist John M. Fabian in place while he sleeps.

Sleeping in Space

Even though they don't work their muscles very hard, astronauts are very tired at the end of their day. Their schedule gives them time to relax and to get eight hours of sleep.

Working together so closely can be hard on astronauts. Sometimes people want a little **privacy**, especially when they sleep. Each astronaut has a tiny, private bedroom the size of a shower stall.

Astronauts don't want to float away as they drift off to sleep. Most astronauts use a sleeping bag or straps to help them stay put while they dream.

Keeping in Touch

Astronauts on board the space station communicate with Earth by using satellite data systems. They can keep in touch with their families by e-mail.

What about those extra-special occasions? Astronaut Robert Curbeam was on board the space shuttle when his daughter turned nine years old. He couldn't be at her birthday party on Earth, so he and the crew of the space shuttle sang "Happy Birthday" to her from outer space.

CHAPTER 5

Exploring Our Solar System

Space exploration can help us understand how our **solar system** was formed. Our solar system includes the sun, the planets that orbit the sun, and the moons that orbit the planets.

Many people wonder about the other planets in our solar system. Are there any living things on those planets? Is Earth the only planet in our solar system that can support life? The space program is trying to answer questions such as these about our solar system.

Solar Sampler

A **nebula** is a cloud of gas and dust. Many scientists think that our solar system formed billions of years ago when a nebula collapsed. After it collapsed, grains of dust formed the planets in our solar system. This same material formed moons, comets, and other objects.

If all the planets are made up of the same material, why do some planets, like Venus, have poisonous atmospheres? Why does Earth support life?

Scientists have ideas about why this happened. They can't be sure, though, until they have a better understanding of what the original nebula was made of.

NASA built a spacecraft called *Genesis* to help find the answers to these questions. *Genesis* will fly toward the sun to collect small **particles** of solar wind. These particles were once part of the sun, but they are now floating in outer space.

Workers lower *Genesis* onto a rocket.

Genesis was launched in August 2001 to collect bits of solar wind. As *Genesis* sends the samples back to Earth, scientists from all over the world can study them.

After testing the solar wind samples, scientists may be much closer to an answer about how our solar system began.

Mars Exploration

Scientists have been studying Mars for a long time. In the 1960s NASA sent a spacecraft named *Mariner* to Mars, Venus, and Mercury. There were no astronauts on board.

Mariner didn't land on the planets. Instead, the spacecraft was designed to stay in outer space and send pictures and information about the planets back to Earth.

Later spacecraft have landed on Mars. From these missions, scientists now know a lot about our closest neighboring planet. There is still much more to learn about Mars.

Sending astronauts to explore Mars is one of NASA's long-term goals. Other goals for Mars exploration include studying the planet's climate and **geology**.

Mars is similar to Earth in many ways. Mars has ice caps at its north and south poles. There are clouds in the Martian sky. Volcanoes and canyons can be seen on the surface of Mars.

There are also many differences between Mars and our own planet. Scientists think there may not be any oxygen in the air on Mars. The polar ice caps on Mars are not made of frozen water as are those on Earth. Instead, they are made of another substance. The sky is pink instead of light blue. The surface of Mars is dry. There are no Martian oceans, but scientists think there may have been water on Mars a very long time ago.

NASA plans to learn whether there ever was life on Mars and whether Mars can support human life in the future.

The surface of Mars appears to be dry and rocky.

Our Future in Space

Space travel existed in science fiction long before the first real rocket blasted off from Earth. Books and movies showed ordinary people living and working on other planets and in space colonies. Tourists in space were also a common theme in science fiction. Space tourists are no longer found only in science fiction stories.

Space Tourist

In April 2001, Dennis Tito, a California **millionaire**, paid the Russian government twenty million dollars for an eight-day trip on board the *International Space Station.*

Tito feels he got his money's worth. He thought his trip into outer space was like a dream come true.

Tito had seen two others go on paid trips into space. In 1990 the Russians sent Tohiro Akiyama, a Japanese television **journalist**, to the *Mir* space station. In 1991 a British **chemist** named Helen Sharman also visited *Mir*.

These two "tourists" did not pay their own way like Tito did. A Japanese television station paid $28 million for Akiyama's one-week stay on board the Russian space station. A British group paid for most of Sharman's trip.

Tito flew to and from the space station on board a Russian Soyuz spacecraft. Two cosmonauts were with him on the Russian rocket. They were on their way to work in the space station.

Dennis Tito already knew a lot about space travel. He had worked for NASA as an engineer. He trained in Russia for many months before his tourist flight.

Dennis Tito's space fare covered the entire cost of launching the Soyuz rocket. Russia may plan more tourist flights to the *ISS*. Some people think the United States should do the same.

DENNIS A.
ДЕННИС А. ТИТО
PK

Dennis Tito proved that the space station could support an extra person. He also showed that an ordinary person could travel in space.

Some people don't like the idea of turning the space station into a hotel for tourists. Those people want the space station to be used only as a science research center.

One person encouraging space tourism is former astronaut Buzz Aldrin. Aldrin pointed out that the space shuttle often flies with only five or six people, even though it can hold seven or eight. He thinks NASA should sell tickets for these seats—at twenty million dollars a ride! That money could go a long way toward paying for more space programs.

Space Colonies

Space tourists are the first ordinary people to go into outer space. There may be a time in the future when many ordinary people will travel in space. Some people may even choose to live in space.

Dennis Tito gets ready to blast off on a Soyuz rocket bound for the *ISS*.

A space colony is a permanent settlement in outer space for ordinary people. No space colonies presently exist.

Many science fiction writers have described space colonies in their books. Scientists are now working on turning this **fantasy** into a real possibility.

Building a space colony is a huge job. It will take a lot of time and money. Modern technology must also advance so there are easier and less costly ways to bring people to outer space. Space station research is giving us a lot of new information about how people can live comfortably and safely in outer space.

Will space colonies be the next case of science fiction becoming fact? Many space scientists think the answer is yes.

Exploring Outer Space

1957

- The Soviet Union launches *Sputnik 1*.

1961

- The Soviets are first to launch a piloted spacecraft.
- Cosmonaut Yuri Gagarin is the first human to orbit the Earth.
- Alan Shepard is the first American astronaut in space.

1962

- John Glenn is the first American to orbit Earth in *Friendship 7*.
- Cosmonaut Valentina Tereshkova is the first woman in space.

1965

- Russian cosmonaut Aleksei Leonov is the first person to "space walk."
- The first American space walk occurs three months later.

1967

- The Soviets launch their *Soyuz 1* rocket.

1969

Apollo 11 astronauts Neil Armstrong and Buzz Aldrin walk on the moon.

1971

The Soviets launch *Salyut 1*, the world's first space station.

1973

Skylab, the first American space station, is launched.

1976

American spacecraft *Viking 1* lands on Mars. It is the first spacecraft to land safely on another planet.

1981

The American space shuttle *Columbia* completes the first piloted flight of a reusable spacecraft.

1983

The first American woman astronaut, Sally Ride, travels in space.

The first African American astronaut, Guion S. Bluford, travels in space.

1990

The *Hubble Space Telescope* is launched from the space shuttle *Discovery*.

1992

Mae Jemison is the first African American woman in space.

1993

Ellen Ochoa is the first Hispanic woman in space, on board space shuttle *Discovery*.

1995

The first phase of the *ISS* begins.

1997

Mars *Pathfinder* lands on Mars.

1998

The first two *ISS* modules are launched and joined in outer space.

2001

The *Genesis* spacecraft begins its three-year mission to collect samples of solar wind.

2002

NASA posts rules about who can visit the *ISS*.

Glossary

achievement (uh CHEEV muhnt) *noun* An achievement is something that somebody has succeeded in doing, usually by hard work.

biofeedback (by oh FEED bak) *noun* Biofeedback is a way to control body processes, such as heart rate or blood pressure, that are not usually under a person's control.

chemist (KEHM ihst) *noun* A chemist is a scientist in the field of chemistry.

complex (kahm PLEHKS) *adjective* Complex means complicated, or not simple.

cooperation (koh ahp uhr AY shuhn) *noun* Cooperation is the act of working together.

cosmonauts (KAHZ muh nawts) *noun* Cosmonauts are astronauts in the Soviet or Russian space program.

dehydrated (dee HY drayt ihd) *adjective* Dehydrated food has had the moisture removed from it.

evaluate (ee VAL yoo ayt) *verb* Evaluate means to judge the value or worth of something.

expandable (ehk SPAND uh buhl) *adjective* Something is expandable if it can be made larger.

fantasy (FANT uh see) *noun* A fantasy is a product of someone's imagination.

full-sized (FUL SYZD) *adjective* Something is full-sized when it is the normal size for its kind.

funnel (FUHN uhl) *noun* A funnel is a tube with a wide end and a narrow end.

geology (jee AHL uh jee) *noun* Geology is the study of a planet's structure.

journalist (JUR nuhl ihst) *noun* A journalist is a writer or a reporter for a newspaper, magazine, or television news show.

magnets (MAG nihts) *noun* Magnets are pieces of metal that attract other metal objects.

mankind (man KYND) *noun* Mankind is all human beings.

manufacture (man yoo FAK chuhr) *verb* To manufacture is to make or to produce.

microwave oven (MY kroh wayv UHV uhn) *noun* A microwave oven is an oven that uses short waves of energy to cook food.

military (MIHL uh tehr ee) *adjective* Military means having to do with armed forces or war.

millionaire (mihl yuh NEHR) *noun* A millionaire is a person who has at least one million dollars.

modules (MAHJ oolz) *noun* Modules are sections or units of a larger thing.

nausea (NAW shuh) *noun* Nausea is a feeling of being sick to one's stomach.

nebula (NEHB yuh luh) *noun* A nebula is a cloud of gas and dust.

overtake (oh vuhr TAYK) *verb* To overtake is to catch up with and pass someone with whom you're racing.

particles (PAHRT ih kuhlz) *noun* Particles are very small pieces of matter.

privacy (PRY vuh see) *noun* Privacy is space or time that a person has for being alone.

proteins (PROH teenz) *noun* Proteins are chemical compounds that occur in plants and animals.

radiation (ray dee AY shuhn) *noun* Radiation is energy that travels in waves or rays.

reusable (ree YOO zuh buhl) *adjective* Something that is reusable can be used again.

roller coaster (ROHL uhr KOHS tuhr) *noun* A roller coaster is an exciting ride found at theme parks and carnivals.

science fiction (SY uhns FIHK shuhn) *noun* Science fiction refers to imaginative stories that are set in the future, in outer space, or on another planet, or that include scientific themes.

significant (sihg NIHF uh kuhnt) *adjective* Significant means important.

simulator (SIHM yoo layt uhr) *noun* A simulator is a training machine that copies real conditions and activities, such as those in a spacecraft.

solar system (SOH luhr SIHS tuhm) *noun* A solar system is a group of bodies that includes a star, such as the sun, and any planets, moons, or other matter that orbit that star.

space shuttle (SPAYS SHUHT uhl) *noun* A space shuttle is a spacecraft that can be used more than once to travel between Earth and space.

technology (tehk NAHL uh jee) *noun* Technology is the science and study of tools and machines and their use.

test pilots (TEHST PY luhts) *noun* Test pilots are people who fly new aircraft to see how the aircraft perform.

weightlessness (WAYT lihs nehs) *noun* Weightlessness is the condition of not having any weight.

Index